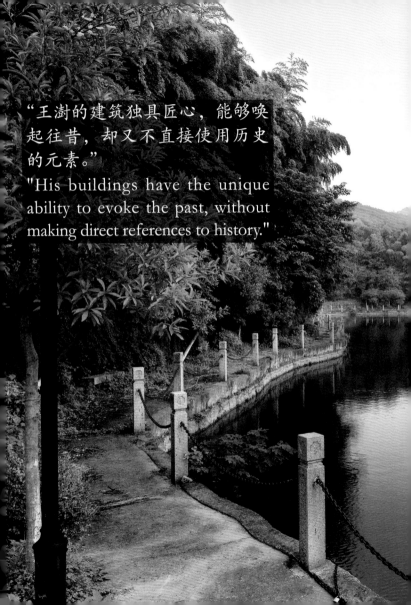

"王澍的建筑独具匠心，能够唤起往昔，却又不直接使用历史的元素。"

"His buildings have the unique ability to evoke the past, without making direct references to history."

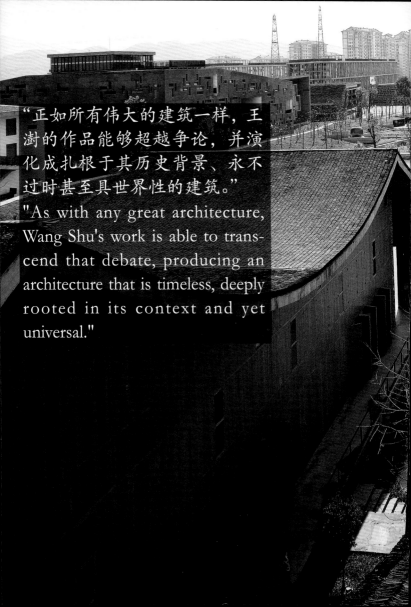

"正如所有伟大的建筑一样，王澍的作品能够超越争论，并演化成扎根于其历史背景、永不过时甚至具世界性的建筑。"
"As with any great architecture, Wang Shu's work is able to transcend that debate, producing an architecture that is timeless, deeply rooted in its context and yet universal."

策划：江岱

文字整理：张翠
书籍设计、地图：孙晓悦
摄影：吕恒中
翻译：江岱

鸣谢：
《时代建筑》杂志社
戴春　段雄春　冯惠民　黄忠顺
江牧　姜庆共　秦蕾

开篇文字引自2012年普利兹克
建筑奖评审辞

Curator: Jiang Dai

Information Sorted: Sarah Zhang
Book Designer, Illustrator: Sun Xiaoyue
Photographer: Lv Hengzhong
Translator: Jiang Dai

Acknowledgements:
Time + Architecture Magazine
Dai Chun, Duan Xiongchun, Feng Huimin
Huang Zhongshun, Jiang Mu
Jiang Qinggong, Qin Lei

Opening text quoted from 2012 Pritzker
Architecture Prize Jury Citation

同济大学出版社
TONGJI UNIVERSITY PRESS

Wang Shu Architecture

王澍建筑地图

同济大学出版社
TONGJI UNIVERSITY PRESS

城市行走编委会编 *CityWalk* Editorial Board

目录

请先阅读	13
王澍的实验建筑	16
走近王澍的建筑	32
——14个王澍建筑旅行指南	
王澍简历	152
作品名录	153
已发表中文文章及著作	156
获奖情况	158
参考阅读	159

Contents

First Stop Reading	13
Wang Shu's Experimental Architecture	16
A Close look at Wang Shu's Architecture	32
— Visitor's Guide to 14 Sites Designed by Wang Shu	
Resume of Wang Shu	152
Chronology of Works	153
Publications in Chinese	156
Awards	158
Reference Reading	159

请先阅读

由王澍设计、建造并保存下来的建筑物目前主要分布在江浙一带,位于上海(1个)、苏州(1个)、南京(1个)、杭州(4个)、宁波(4个)、海宁(1个)、金华(1个)等7个城市;另广州东莞也有一个。

这些建筑物的营造时间主要集中在21世纪的前十年,最早产生影响力的是苏州大学文正学院图书馆,之后有杭州的中国美术学院象山校区一期、二期、"钱江时代"垂直院宅,以及宁波的宁波美术馆、宁波博物馆等代表性建筑群和建筑,整体体现出王澍"用这个世纪的一切现代语言对过去和现代的建筑作品进行试验的探索"的创作取向。

First Stop Reading

The architecture designed and built by Wang Shu, which are preserved till now, are mainly spread out in the region of Jiangsu & Zhejiang Provinces, including 7 cities: Shanghai (1), Suzhou (1), Nanjing (1), Hangzhou(4), Ningbo(4), Haining(1), and Jinhua(1). There is another work built in Dongguan, a city near Guangzhou (Canton, China).

Most of these works were built in the first decade of 21st century. The earliest significant impact arose with his work at Library of Wenzheng College in Suzhou University. After that, represented by Xiangshan Campus of China Academy of Art (Phase I & II), Vertical Courtyard Apartments — Qianjiang Epoch, together with Ningbo Contemporary Art Museum, Ningbo History Museum etc., Wang Shu's ideas and philosophy in his architecture gradually got together. In his own words, that's: Utilize all modern architecture language of the century to test past old and current new architecture.

2 苏州大学文正学院图书馆
Library of Wenzheng College, Suzhou University (1999-2000)

编号、建筑名称、设计建造时间
Index #, Project/Work name, Period of design & construction

"如何让人生活在处于'山'和'水'之间的建筑中，以及苏州园林的造园思想是我设计这座图书馆的沉思背景。"

"The thought that how the daily living would be like in an architecture melting into *Shan-Shui* ('*Shan-Shui*' means 'High mountain and flowing water' in Chinese, which is common icon used to represent the whole natural environments in traditional Chinese culture) surroundings, and the philosophy in garden construction of Suzhou's, are the templates that I ponder into as I work on this library project."

引文
Quotation

王澍最早在建筑设计界奠定地位的作品之一。

One of the earliest works of Wang Shu, which establish his reputation in his professional career.

简介
Briefing

房子的基本状态是"睡着了"。

产自苏州的水磨清砖贴面，以及切割打磨的大青砖。

Basically speaking, this house is in its sleep.

The levigated blue brick veneers made in Suzhou, together with large blue brick after cutting and polishing.

图片说明
Picture Illustrations

我们以普通人的视角拍摄和记录王澍建筑这一特别的公共存在，不代表任何官方或专业立场。

本书所收信息截止至 2012 年 6 月。书中的建筑名称、地址和建造年代，均以 2012 年普利兹克建筑奖公布的媒体资料为主要依据。引文及图片说明文字全部摘自王澍历年来发表在各大出版物上的文章。

Adopting perspectives of ordinary people, we try to make visual and written records of Wang Shu's architecture as a special public existence. There is no indication of any kind that this represents official or professional position taken by any specific person or entity.

All contents and materials here are collected from public resources available by June, 2012. The work or project names, locations and construction periods etc. are quoted from media kit provided by 2012 Pritzker Prize Committee. Quotations and picture illustrations are quoted from Wang Shu's publications.

王澍的实验建筑
Wang Shu's Experimental Architecture

"每一次,我都不只是做一组建筑,每一次,我都是在建造一个世界。"

"Each and every time, I worked not only on single individual, or single group of architecture; each and every time, I was building a whole world."

五散房
Five Scattered Houses

"业余建筑工作室的基本工作方式,从田野调查入手,和一组地方工匠长期配合,由小型建造实验开始,逐渐形成大型建筑的设计与施工方法。"

"As in the basic work agenda in our Amateur Architecture Studio, we start from field survey, build up long term work relations with a group of local craftsmen. After initiating small building experiments, we then gradually accumulate experiences and common work routines in designs and constructions of large projects (in this area)."

施工现场
Construction site

"我们在宁波发现了砖的特殊砌法。回收旧料在那个地区非常普遍，所有的墙体都是杂砖，10平方米里面能数出84～87种不同的砖，这种技术发展到了如此炉火纯青的地步。"

"In Ningbo, we have found a unique brickwork technique. Recycling construction materials is quite popular in this region. Nearly every wall is built with a hybrid of different makes of bricks, you can count up to 84-87 different brick types within an area of 10 square meters on a wall — just imagine what an artistic state this technique has been grown into."

宁波博物馆
Ningbo History Museum

"2006年夏,业余建筑工作室的5个同事、与我们多年共事的3个工匠和我一行9人去威尼斯建造'瓦园'。决定做什么并不难,难的是如何做。"

"……我就跟大家说,要按《营造法式》道理去做。……'瓦园'最终只用13天建成。"

"……记得双年展技术总负责雷纳托来检查,他在'瓦园'的竹桥上走了几个来回,诚挚地告诉我:真是好活。但有意思的是,他的眼中没有看到什么'中国传统',而是感谢我们为威尼斯量身定做了一件作品,他觉得那大片瓦面如同一面镜子,如同威尼斯的海水,映照着建筑、天空和树木。"

"他肯定不知道,我决定做'瓦园'时曾想到五代董源的'水意'。'瓦园'最后如我所料,如同匍匐在那里的活的躯体,这才是'营造'的本意。"

"In summer of 2006, together with 3 craftsmen who have been working with us for years, 5 colleagues from Amateur Architecture Studio and I went to Venice to build Tiled Garden. The difficult part is not what to build, but how to build it."

"... Then I told them, we'd do it in the way described in *Yingzao Fashi* (Methods and standards in constructions, a book from Song Dynasty) . ... It took only 13 days for us to complete the work of Tiled Garden."

"... I can still remember Mr. Renato Russi, who is the head of construction supervision committee of the Biennale, came to do the final walk through. He walked back and forth several times on the bamboo bridge in the garden, then turned to me, 'seriously, a nice job.' But interesting enough, he saw no so-called 'Chinese tradition' in his eyes, but just gratefulness to us for tailoring a perfect cut for Venice — the huge area of tiles, more or less as the huge water surface of Adratic Sea, reflecting architecture, sky and trees of Venice."

"He probably never imagined that the 'water sensation' coined by Dong Yuan in Five Dynasties Era is one of the original initiatives of this very design. Exactly as I prospected, the Tiled Garden turned out to be a living Creature crouching on the ground — that should be the original ideal realm of building conception itself."

"关于造园,近两年我常从元代画家倪瓒的《容膝斋图》讲起……《容膝斋图》的意思,就是如果人可以生活在如画界内的场景中,画家宁可让房子小到只能放下自己的膝盖。"

"如果说,造房子,就是造一个小世界,那么我以为,这张画边界内的全部东西,就是园林这种建筑学的全部内容,而不是像西人的观点那样,造了房子,再配以所谓景观。"

"Talking of garden construction, I often start from the painting of *Rongxizhai Tu* painted by Ni Zan in Yuan Dynasty... The title suggests, if one could live within the set-up that the painting has described, the painter would rather have his house as small as his knees."

"In the sense that building one house means building a small world, everything within that paint frame means to be everything inside the subject of whole garden building—as one genre of architecture. This is different from the views of western culture: they build the house first, and then match it with so-called landscape."

元人倪瓒《容膝斋图》
Rongxizhai Tu by Ni Zan, Yuan Dynasty

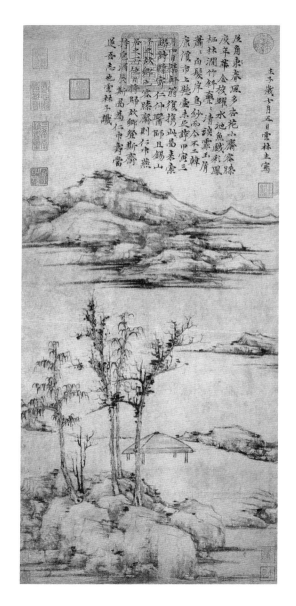

宋人李唐《万壑松风图》
与宁波博物馆
Wanhesongfeng Tu (Pine Wind from Myriad Villages) by Li Tang, Song Dynasty vs Ningbo History Museum

后页:宋人王希孟《千里江山图》
与象山校区建筑
Next Page: *Qianlijiangshan Tu* (Landscape of a Thousand Miles) by Wang Ximeng, Song Dynasty vs Xiangshan Campus architecture

走近王澍的建筑
—— 14个王澍建筑旅行指南

A Close Look at Wang Shu's Architecture
— Visitor's Guide of 14 Sites Designed by Wang Shu

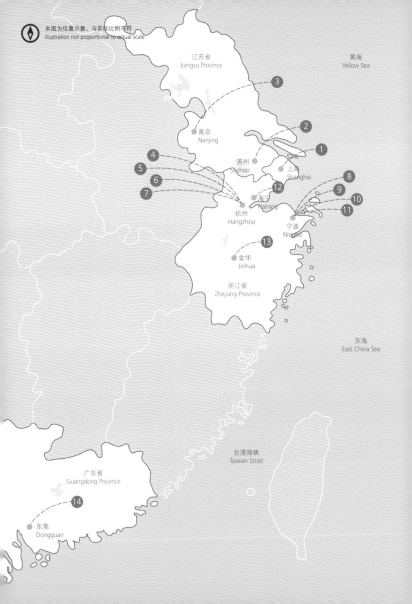

1. 上海世博会宁波滕头馆 37, 38
2. 苏州大学文正学院图书馆 37, 42
3. 三合宅 37, 48
4. 垂直院宅 55, 56
5. 中国美术学院象山校区 55, 60
6. 中山路旧街区综合保护与更新 55, 88
7. 南宋御街博物馆 55, 94
8. 宁波美术馆 99, 100
9. 五散房 99, 108
10. 华茂外国语学校华茂美术馆 99, 114
11. 宁波博物馆 99, 122
12. 海宁青少年中心 135, 136
13. 瓷屋 135, 140
14. 东莞理工学院音乐与舞蹈系教学楼 135, 146

1. Ningbo Tengtou Pavilion, Shanghai Expo............37, 38
2. Library of Wenzheng College, Suzhou University..37, 42
3. Sanhe House 37, 48
4. Vertical Courtyard Apartments 55, 56
5. Xiangshan Campus, China Academy of Art 55, 60
6. Old Town Conservation and Intervention of
   Zhongshan Street 55, 88
7. Exhibition Hall of the Imperial Street of
   the Southern Song Dynasty 55, 94
8. Ningbo Contemporary Art Museum 99, 100
9. Five Scattered Houses 99, 108
10. Huamao Art Museum, Huamao Foreign
    Language School 99, 114
11. Ningbo History Museum 99, 122
12. Youth Center, Haining 135, 136
13. Ceramic House 135, 140
14. Teaching Building of Music and Dance Department,
    Dongguan Institute of Technology 135, 146

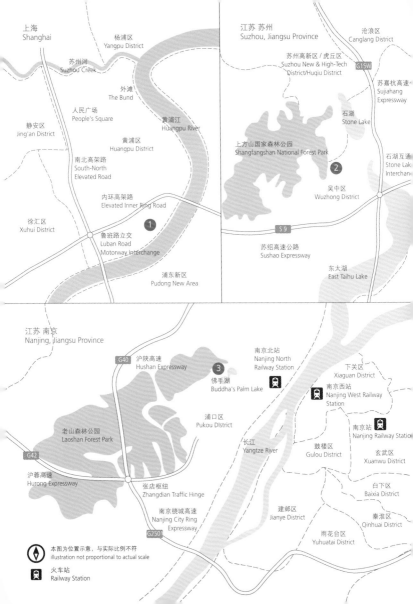

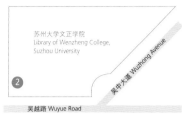

1. 上海世博会宁波滕头馆
   上海市黄浦区浦西世博园 E 片区
   中山南路（近保屯路）
   38 ~ 41 页
1. Ningbo Tengtou Pavilion, Shanghai Expo Zone E, Puxi Expo Site, Huangpu District, Shanghai
   South Zhongshan Road (near Baotun Road)
   pp38-41

2. 苏州大学文正学院图书馆
   苏州市吴中区吴中大道 1188 号
   苏州大学文正学院内
   42 ~ 47 页
2. Library of Wenzheng College, Suzhou University
   1188 Wuzhong Avenue, Wuzhong District, Suzhou In the Wenzheng College, Suzhou University
   pp42-47

3. 三合宅
   南京市浦口区岔口路 9000 号
   老山森林公园佛手湖景区内，四方艺术湖区
   48 ~ 53 页
3. Sanhe House
   9000 Chakou Road, Pukou District, Nanjing
   Sifang Collective, in the Buddha's Palm Lake Scenic Area of Laoshan Forest Park
   pp48-53

1 上海世博会宁波滕头馆
Ningbo Tengtou Pavilion, Shanghai Expo
(2010)

"只表现滕头的现在,我不知道该如何做,但我有兴趣做一个建筑,剖切一下这个地区乡村建筑过去与现在的差别,也许能推断一下它的未来。"

"I am not quite sure what to say to present Tengtou. But no doubt that I would like to take a section look across past and present of the village architecture in this region. Maybe this would help when we speculate its future."

2010年上海世博会上,王澍让他的"瓦爿墙"影响力进一步扩大,试图通过滕头描绘一种新乡村的全景。世博会结束后场馆计划移回宁波滕头村。

In 2010 Shanghai World Expo, Wang Shu expanded the impact of "wall construction with recycled bricks and tiles technique" with his continuing promotion. He has painted the panorama of a new village living. The pavillion would be on schedule moved back to Tengtou Village, Ningbo after the show.

门洞就是一个洞，进到洞内，只能从剖面一层层去观看。

主要材料是竹条模板混凝土、回收旧砖瓦和竹材。

The entrance is nothing but an opening, once entering, you can only observe through its sections, layer by layer.

Major materials are bamboo molded concrete, recycled bricks and tiles, and bamboos.

2 苏州大学文正学院图书馆
Library of Wenzheng College, Suzhou University (1999-2000)

"如何让人生活在处于'山'和'水'之间的建筑中，以及苏州园林的造园思想是我设计这座图书馆的沉思背景。"

"The thought that how the daily living would be like in an architecture melting into *Shan-Shui* (*'Shan-Shui'* means 'High mountain and flowing water' in Chinese, which is the common icon used to represent the whole natural environments in traditional Chinese culture) surroundings, and the philosophy in garden construction of Suzhou's, are the templates that I ponder into as I work on this library project."

王澍最早在建筑设计界奠定地位的作品之一。

One of the earliest works of Wang Shu, which establish his reputation in his professional career.

按照造园传统，建筑在"山水"之间最不应突出，这座图书馆将近一半的体积处理成半地下，从北面看，三层的建筑只有二层。

水中那座亭子般的房子……便是一个中国文人看待所处世界的"观点"。

四个散落的小房子和主体建筑相比，尺度悬殊，但在这里，可以相互转化的尺度是中国传统造园术的精髓。

In the context of garden construction tradition, the last thing you want would be having an "out–standing" architecture located within the *Shan-Shui* surroundings. Nearly half of the library body is deliberately embedded below the ground. The three storey building suggests only two levels on its north side.

Standing above water, this pavilion-like house is more or less the "viewpoint" of a traditional Chinese literate when he looks into the world surrounding him.

Comparing with the main building, the four small scattered houses are of great difference in scale and dimensions. But exactly as in traditional Chinese garden set-up, such ever inter-transferable scaling fractals are part of the fundamental essences and spirits of its philosophy.

3 三合宅
Sanhe House (2003)

"这个三面围合一面开敞的建筑,在空间上是内聚和封闭性的,在形态上保持建筑与空间的连续性。这种连续性不仅在于建筑本身,也体现在建筑与城市的关系上,是设计者对于'中国房子'范形的一次具体的操作。"

"This is a house with three sides closed and one side open. The space flow here is inherently coherent and enclosing. Nothing but the formation of the architecture holds things together and keeps the continuity of spatial experience. Such experiences extend far beyond this architecture itself by projecting much insight into the connection between a large city and the architecture within. This is a real-world building practice of 'Chinese House' model from the architect."

四方当代艺术湖区依山而建,2003 年始邀请了来自 13 个国家的 24 位大师级建筑师每人设计一个建筑作品。王澍的三合宅就位于山顶。这一地块上原本生长着两棵合抱的黄连树。

Laid along the hillside in Sifang Contemporary Art Lakeshore, works by 24 master architects from 13 countries, each for one, were set up here in 2003. The Sanhe House designed by Wang Shu was located near the hilltop, where originally stand two goldthread trees that two man could encircle each truck with his stretched arms.

房子的基本状态是"睡着了"。

产自苏州的水磨清砖贴面,以及切割打磨的大青砖。

Basically speaking, this house is in its sleep.

The levigated blue brick veneers made in Suzhou, together with large blue brick after cutting and polishing.

4. 垂直院宅（钱江时代）
   杭州市上城区清江路 346 号
   56 ~ 59 页
4. Vertical Courtyard Apartments
   (Qianjiang Epoch)
   346 Qingjiang Road, Shangcheng
   District, Hangzhou
   pp56-59

5. 中国美术学院象山校区
   杭州市西湖区转塘镇象山路 352 号
   60 ~ 87 页
5. Xiangshan Campus, China
   Academy of Art
   352 Xiangshan Road, Zhuantang Town,
   West Lake District, Hangzhou
   pp60-87

6. 中山路旧街区综合保护与更新
   杭州市上城区中山路鼓楼至西湖大道段
   88 ~ 93 页
6. Old Town Conservation and Intervention
   of Zhongshan Street
   Along Zhongshan Street, from
   Gu Lou (Drum Tower) to West Lake Blvd.,
   Shangcheng District, Hangzhou
   pp88-93

7. 南宋御街博物馆
   杭州市上城区中山路鼓楼至西湖大道段
   94 ~ 97 页
7. Exhibition Hall of the Imperial Street of
   the Southern Song Dynasty
   From Gu Lou (Drum Tower) to West Lake
   Blvd., Shangcheng District, Hangzhou
   pp94-97

4 垂直院宅
Vertical Courtyard Apartments
(2002-2007)

"我坚持认为,中国的实验建筑活动如果不在城市中最大的建设活动——住宅中展开实践,那么它将是自恋而且苍白的。"

"I still insist, the experimental architecture of China would be narcissistic and weak, if its experiments never get involved in the largest metro construction activities—the metro residential architecture."

从机场进入杭州,越过钱塘江第三大桥,引桥左侧,看到的第一组建筑。王澍的城市住宅实验建筑。

On your way from Xiaoshan airport to downtown Hangzhou, after passing the Third Bridge on the Qiantang River, the first group of buildings on the left of the approach bridge are Vertical Courtyard Apartments—Wang Shu's experimental metro residential architecture.

哪怕住在100米的高度上,也能体会到住在两层高的小楼里的感觉,屋檐滴雨,窗前有树。

这已经不是普通的住宅设计,而是在知道不能回避规定性的同时,实验一种能容纳自发性的城市居住方式,显示一种对土地的眷恋,验证一种生活世界的理想。

在设计城市住宅的同时思考对城市的超越。

Even residing in the air of 100 meters high, one would still be able to be touched by the surroundings of two-level lodge house: raindrops dripping off the roof edge, trees growing in front of the windows.

This has been far from an ordinary residential design. This is an experiment of metro living framework that can embrace those natural and spontaneous living behavior of human beings, providing madatory restrictions that you can not bypass. This is an expression of human longing for the earth. This is a test of living ideas in real world.

Rethink the future of metro living as we work on the design of metro residential architecture.

5 中国美术学院象山校区
Xiangshan Campus, China Academy of Art

"象山校区主要是做一个关于具有中国本土特点现代建筑的一种实验,灵感来自中国传统的山水绘画和自然相互对话的观念。"

"Most of Xiangshan campus design is about an experiment to build modern architecture with local characters from China. They are inspired by Chinese traditional Shan-Shui paintings and the idea of communications between human and the nature."

象山校区是王澍实践本土建筑学的最大作品。在设计过程中,为充分发掘建筑材料的可再利用和经济适用性,王澍从各地的拆房现场收集了700多万块不同年代的旧砖弃瓦,让它们在象山校区的屋顶和墙面上重现新生。

Xiangshan Campus of China Academy of Art is the largest project that Wang Shu has worked on in his practice of "local architecture". Through the whole designing process, Wang Shu collected over 7 million used and deserted bricks or tiles with different manufacture periods from demolishing sites all over China, and put them into new lives on the roofs and walls of Xiangshan Campus. This is his practice to explore the full recycling lives of building materials and his ideas of economical construction.

一期工程
Phase I (2002-2004)

象山一期是由 10 座主体建筑构成的建筑群体，包括 1 座图书馆、1 座小美术馆、6 座教学和作坊综合楼，1 座小体育馆，1 座工作室及管理塔楼。另有连接建筑和山体的桥梁两座，长度分别为 88 米和 180 米。

The phase I of Xiangshan Campus project consists of 10 main buildings, including 1 library, 1 small gallery, 6 classroom & workshop complexes, 1 small gym and 1 studio & administration tower. Besides, there are two bridges, 180 meters and 88 meters long respectively, connecting the buildings and the adjacent hills.

从山上往下看,整个一片瓦的世界。

中国的山与建筑的关系,从来不是景观关系,而是某种共存关系。

校园建筑最终落实为一种"大合院"的母题聚落,一座消瘦的玻璃塔被放在精心选择的位置,形成"面山而营"的"塔院式"格局。

山边原有的溪流、土坝、鱼塘均被原装保留,只做简单休整。清淤产生的泥土用于建筑边的人工覆土,溪塘边的芦苇被复种。

Seen from the hilltop, it is a world of tiles.

Concerning the relationship of the mountains or hills to architectures nearby in Chinese tradition, it's never just part of the scenery; more or less they co-exist with esch other.

In the end, the campus architecture design has evolved into a motif collection as "the gathering courtyard". A slim cut glass-tiled tower is located at a deliberately selected position, gradually, the formation model of a tower plus a courtyard facing the hill comes into being.

The creeks, original soil dams and ponds are all conserved as they were except absolutely necessary fixing. Mud removed from the construction sites is all used as surface cover of the backfill. Reeds are replanted by the waterside.

二期工程
Phase II (2004-2007)

二期工程由 10 座大型建筑与两座小型建筑组成，包括建筑艺术学院、设计艺术学院、实验加工中心、美术馆、体育馆、学生宿舍与食堂。

The Phase II of Xiangshan Campus project consists of 10 large buildings and 2 smaller ones, including Department of Architecture, the Professional School of Art Design, Experimental & Machining Center, art gallery, gym, student dormitory and canteen.

象山校区二期建筑群有一个原则,就是建筑的高度大多要比树低一点。整个校园都在以某种方式顺应这里的自然原则。

校园建筑在"自然"与"城市"之间的思考中显现出来。

One principle was observed in designing process of the Phase II architecture, that is the height of most buildings should be shy to the trees, as a gesture to acknowledge and adopt the natural rules here.

The architecture on the campus emerge from the thinking of the interactions between the "nature" and the "city".

6 中山路旧街区综合保护与更新
Old Town Conservation and Intervention of Zhongshan Street (2007-2009)

"我所热爱的中山路,和所有我喜欢的中国城市街道一样,新旧混杂,逸趣横生,人们在门前、窗下、街角街边随时发明着各种建筑的用法。"

"Same as all the Chinese city streets that I am fond of, Zhongshan Rd. beloved by myself, is a mixing template of fashions and antiques, filled with excitement and fun. Anytime and anywhere, people are practicing their own renovating uses of the architectures here."

受杭州政府的委托,王澍设计并主持修建了市内中山路从鼓楼到西湖大道1公里段的步行街。

Appointed by Hangzhou Municipal Government, Wang Shu presides the designs and renovations of the pedestrian street about 1 km long along Zhongshan Rd. It extends from Gu Lou (Drum Tower) to West Lake Blvd. in downtown Hangzhou.

在很多城市，一些街道原本拥有历史上不同时代的各种痕迹，都有其独特的城市记忆。

100 年前，这条街曾经是杭州最繁盛的街道；也是杭州宗教最集中、最早出现西方建筑的街道。

黑色、灰色和白色应该是最具代表性的杭州颜色，因为自然界的颜色在这三种颜色当中是最能突显的。

In many cities, streets gradually possess a variety of touches from different times in history with unique memories.

100 years ago, this was the most prosperous street in Hangzhou, with the highest concentration of diverse religious architectures, and it saw the very first western architecture in Hangzhou.

Black, grey and white are supposed to be the colors speaking for Hangzhou, and colors of the nature would be most catchy in the background of them.

7 南宋御街博物馆
Exhibition Hall of the Imperial Street of the Southern Song Dynasty (2007-2009)

"整个建筑为一个木构瓦面的多折大棚覆盖,它对中山路遗迹周围建筑都呈开放状态。……人们也可以把这里用作只是穿越城市迷宫的过道,建筑因此而真正嵌入城市生活之中。"

"The whole building is covered by a tiled multi-folded roof of wood structure. It takes an open gesture to embrace surrounding architectures in Zhongshan Rd. district. ...As it could be used by the public as a gateway to go through the city street maze, this architecture therefore got blended into the daily urban living of the very city."

御街博物馆为展示埋在杭州中山路旧街区最底层、800年前的南宋御街而设。

Exhibition Hall of the Imperial (Yu Jie) Street was set up to exhibit the relic layer of Zhongshan Rd. street blocks. 800 years ago, it's the Imperial Street of Southern Song Dynasty.

设计借鉴了浙江南部古廊的编木拱结构桥。

馆内功能分为展览和茶饮,它们之间的关系似有似无。

The design borrow ideas from the old woven timber arch bridge (lounge bridge) existing in south Zhejiang Province.

There are two function zones inside the building: exhibition and cafe, which are loosely connected.

8. 宁波美术馆
   宁波市江北区人民路 122 号
   100 ~ 107 页
8. Ningbo Contemporary Art Museum
   122 Renmin Road, Jiangbei District, Ningbo
   pp100-107

9. 五散房
   宁波市鄞州区鄞州公园内（近首南中路）
   108 ~ 113 页
9. Five Scattered Houses
   In the Yinzhou Park (near Middle Shounan Road), Yinzhou District, Ningbo
   pp108-113

10. 华茂外国语学校华茂美术馆
    宁波市鄞州区鄞县大道中段 2 号
    华茂外国语学校内
    114 ~ 121 页
10. Huamao Art Museum, Huamao Foreign Language School
    2 Yinxian Avenue(M), Yinzhou District, Ningbo
    In the Huamao Foreign Language School
    pp114-121

11. 宁波博物馆
    宁波市鄞州区首南中路 1000 号
    122 ~ 133 页
11. Ningbo History Museum
    1000 Middle Shounan Road, Yinzhou District, Ningbo
    pp122-133

8 宁波美术馆
Ningbo Contemporary Art Museum
(2001-2005)

"在我看来，比一座美术馆的样式更重要的是先在的城市结构，而城市的记忆，应包含到今天为止所发生的一切事件线索。"

"In my opinion, the existing urban texture is more important than the form of a museum, while the urban memory should include the threads of all happenings that have ever taken place here by now as a total."

位于宁波外滩边缘，由原宁波港轮船码头候船大楼重建而成。建筑契合到城市结构中去的实践之一。

Located on the edge of Ningbo bund, the museum is a renovation from original boarding terminal hall of passenger ferries which belongs to Ningbo Ports. It is one of the practices to weave the architecture into the urban texture.

两个扁平长方形互相平行，横卧江边，再次向甬江暗示和重构了码头与船的语句结构。面向城市则暗示和重建了传统中国城市院落与城市的结构。

Two flat rectangles laid along the bund, parallel to each other. Once again, the indication of port & ferry idea is revisited at Yong River bund. As it's facing the city, the traditional configuration of Chinese city and the urban courtyard within is revisited here too.

建筑表皮材料的使用同样暗示了城市记忆线索的混合性。基座青砖是传统宁波的建筑主材,而上部钢木材料则是船与港口的建造主材。

沿江青砖基座的洞窟直接取材于敦煌片段,指出这里曾是宁波人去普陀进香的出发之地。

The mixture of surface materials on the architecture also suggests that of the urban memory threads. The blue bricks around the architecture base is adopted as the traditional major construction material of Ningbo, meanwhile wood and steel above the base are selected since they are dominant construction materials of ferry ships and the port itself.

The caved-in holes appears on the blue brick base are motifs quoted directly from Dunhuang – which indicates it's once the boarding dock of people of Ningbo on their (Budda) worship trip to Putuo Mountain.

9 五散房
Five Scattered Houses
(2003-2006)

"用 400 平方米的一座画廊建筑第一次把旧料回收、循环建造的做法实现。这种做法实际上也来自宁波民间。"

"For the very first time, we recycled the building materials from a 400 square meter gallery building and put them back into use. The very idea in fact comes from the local practice in tradition."

王澍在中国美术学院象山校区二期工程前进行的建筑实验,为建设中的鄞州公园设计5幢具有宁波地方特征的小建筑(分散在园中)。王澍这时已掌握了"瓦爿墙"的施工技术,对瓦片的使用也更加纯熟。

It is an architecture experiment performed by Wang Shu before he rolled out his works of Phase II, Xiangshan campus of China Academy of Art. He designed 5 small houses with full Ningbo features in Yinzhou Park (scattered in the park). Wang Shu has mastered the techniques of wall construction with recycled bricks and tiles, and become skillful in utilizing these (recycled) tiles.

正是情趣，在一开始决定你所建造的世界是平衡还是躁动，是深邃还是粗浅，是静寂还是喧嚣。

It's nothing but taste and interest, that decide at the very beginning whether the world you are building is balanced or biased, insightful or shallow, peaceful or irritated.

10 华茂外国语学校华茂美术馆
Huamao Art Museum, Huamao Foreign Language School (2008)

"我们不是设计一个房子,而是要建造一个世界。只有植物、动物、河流、湖泊与人和谐共生才能称为一个美丽的世界。"

"What we are doing here is not designing a house, but a whole world. A beautiful world only happens when plants, animals, rivers, ponds and human could sustainably co-exist in harmony."

全国第一家民办教育校园美术馆,免费向公众开放。体现王澍特有的建筑风格。

The first campus art museum in private educational institution, which reflects Wang Shu's unique architecture style. The admission is free to general public.

中国的建筑从来不是一个物体，最好的感觉总是想让你进到里头去。

Never ever, was chinese architecture a solitary object. As the best part, it always seduces you to walk inside.

11 宁波博物馆
Ningbo History Museum
(2003-2008)

"我想告诉人们,曾经的城市生活是怎样的。10多年前,这是一个美丽的海港城市,有30多个传统村落。到今天,几乎所有的东西都被拆除了,这里变成了一片几乎没有回忆的城市。我把能在这个地区收集到的各种旧建筑材料再次利用,与新材料一起在新的建筑上混合建造。我想建造一个有自我生命的小城市,它能重新唤醒这个城市的记忆。"

"All I want to tell is, what the urban living used to be here. About 10 years ago, this was a beautiful harbor city, possessing more than 30 traditional natural villages. But by now, nearly everything has been dismantled. It has become a city without almost any memory here. What I was trying is to recycle varieties of used building materials that I could collect, accomplish a blended or mixing new constructions out from new and recycled materials side by side. I want to build a small town with its own life, which could once again, wake up the latent memory of the city."

王澍通过国际竞标获得了这个项目,把它作为"重建一种当代中国本土建筑学"的探索和实践。他在上面运用了已经成熟的"瓦爿墙"技术。普利兹克评奖委员会认为,这是最代表王澍的思想和工作特质的作品。

Wang Shu won this project in an international competition. This is a practice and exploration of his idea — To re-invent a local architectural methodology in contemporary China. According to 2012 Pritzker Prize Jury, it's the icon work of Wang Shu's which best defined his ideas and his unique styles in architectures.

作为山的物性是它唯一要表达的。

这类砖、瓦、陶片都是自然材料，是会呼吸的，是"活"的，容易和草木自然结合，产生一种和谐沉静的气氛。

北段浸在人工开掘的水池中，土岸，植芦苇。水有走势，在中段入口处溢过一道石坝，结束在大片鹅卵石滩中。

在建筑开裂的上部，隐藏着一片开阔的平台，通过4个形状不同的裂口，远望着城市和远方的稻田和山脉。

Its materiality is the only thing that mountain intends to say.

Bricks, tiles and pottery plates, as natural materials, could literally breathe. They are "living" materials, could easily melt into nature, and promote the air of harmony and peace.

The north part submerges into a man-made water pond, soil banked, reeds growing. The water gets its way down, overflows a stone dam near the middle section, and ends up dissolving over an open cobble beach.

A wide platform is concealed behind those huge openings on the top half of the elevations, from where one can overlook the metro area, distant crop fields and mountain ranges through the four differently shaped openings.

12. 海宁青少年中心（海宁青少年宫）
    海宁市硖石镇水月亭西路 262 号
    136 ～ 139 页
12. Youth Center, Haining
    262 West Shuiyueting Road, Xiashi Town, Haining
    pp136-139

14. 东莞理工学院音乐与舞蹈系教学楼
    （今师范学院）
    东莞市松山湖大学路，东莞理工学院内
    146 ～ 151 页
14. Teaching Building of Music and Dance Department (Teachers' College now), Dongguan Institute of Technology
    Daxue Road, Songshan Lake, Dongguan
    In the Dongguan Institute of Technology
    pp146-151

13. 瓷屋
    金华市清照路，金华建筑艺术公园内第九号建筑
    140 ～ 145 页
13. Ceramic House
    Qingzhao Road, No.9 architecture in the Jinhua Architecture Art Park, Jinhua
    pp140-145

12 海宁青少年中心
Youth Center, Haining
（1989-1990）

王澍第一个独立设计的建筑项目。

This is the very first architecture project independently designed by Wang Shu.

13 瓷屋
Ceramic House
(2003-2006)

"设计的出发点是要把这个100平方米的咖啡屋作成一个可以盛装风和水的器物。房子取自宋代手砚器型，单层，砚首在南，砚尾在北。"

"The jump start of this project is to make this 100 square meter cafe & tea house an ink stone filled with wind and water. The architecture form was adopted from a handy ink stone configuration, single level, south-north orientation, with head of ink stone (ink side) extending to the south, while tail (grinding side) to the north."

"瓷屋"茶室是对江南院落、风雨诗意和陶瓷运用的全新演绎，位于金华武义江边、艾未未主持的金华建筑艺术公园中。

Utilizing ceramics, this very cafe & tea house re-explains the poetic realm of nature variation in a south China courtyard. It's located in the Jinhua Architecture Art Park along Wuyi riverside in Jinhua, Zhejiang Province. Ai Weiwei presides and coordinates the designs of the very park.

屋内外均贴中国美术学院陶瓷系周武老师做的瓷片,房子就成了彩色的。

东南风吹过,风沿砚坡爬向西北。金华一带多雨,雨沿砚坡自西北下泻东南。

The tiles crafted by Mr. Zhou Wu of ceramic Dept. from China Academy of Art wrap up the house inside out, and turn it into a colorful building.

In seasons that southeast wind dominates, the wind climbs up the slope to the northwest. As a highly rainy area, when it rains in Jinhua, water runs down the roof from northwest to southeast.

14 东莞理工学院音乐与舞蹈系教学楼
Teaching Building of Music and Dance Department, Dongguan Institute of Technology (2003-2005)

王澍参与崔恺等建筑师主持的项目,为理工学院造的一所建筑。

Wang Shu joins Mr. Cui Kai and other architects in this project, designing a building for this univ. of science & technology.

形体方正，简单平静。

Regular form, simple and peaceful.

王澍简历

建筑师 / 教授

出生：中国新疆乌鲁木齐 1963

教育背景：

南京工学院（现东南大学）建筑系建筑学专业本科毕业 1985

南京工学院（现东南大学）建筑研究所建筑硕士研究生毕业 1988

同济大学建筑与城市规划学院建筑学专业博士毕业 2000

工作经历：

进入浙江美术学院研究所从事研究和实践 1988
创办"业余建筑工作室" 1997
开始在中国美术学院任教 2000
出任中国美术学院建筑艺术学院院长至今 2007

Resume of Wang Shu

Architect/Professor

Born: 1963 in Urumqi, Xinjiang Province, China

Education:

Nanjing Institute of Technology,

Department of Architecture, Bachelor of Science in Architecture, 1985

Master Degree in Architecture, 1988

Tongji University, College of Architecture and Urban Planning, Ph.D. in Architecture, 2000

Career:

Worked for the Zhejiang Academy of Fine Arts in Hangzhou doing research on the environment and architecture in relation to the renovation of old buildings, starting from 1988

Founded "Amateur Architecture Studio", in 1997

Became a professor at China Academy of Art in Hangzhou, in 2000

Named head of the Architecture Department, China Academy of Art, in 2007

作品名录

建筑

1985–1987 华侨大厦，中国南京

1989–1990 海宁青少年中心，中国海宁

1991 浙江美术学院国际画廊（建成，已拆除），中国杭州

1991 孤山艺术家沙龙（建成，已拆除），中国杭州

1991 丰乐桥地下通道（建成，已拆除），中国杭州

1993 中国美术学院湖滨校区改造规划（方案）

1993 中国美术学院雕塑系馆（方案）

1994 中国杭州虎跑禅心茶道园（方案）

1996–1999 一个退休建筑教师的住宅（方案）

1997 自宅（室内），中国杭州

1998 陈默艺术工作室（室内），中国海宁

1999–2000 苏州大学文正学院图书馆，中国苏州

1999–2000 顶层画廊（建成，已拆除），中国上海

2000 墙门（装置），中国杭州（已拆除）

2001 一分为二（装置），中国杭州（已拆除）

2001–2005 宁波美术馆，中国宁波

2002–2003 路桥古镇保护、古建修复及协调区（方案），中国台州

2002–2004 中国美术学院象山校区一期，中国杭州

2002–2007 垂直院宅（钱江时代），中国杭州

Chronology of Works

Architecture

1985-1987 Overseas Chinese Building, Nanjing, China

1989-1990 Youth Center, Haining, China

1991 International Gallery of Zhejiang Academy of Fine Arts (completed and demolished), Hangzhou, China

1991 Artist Salon in Gushan Hill (completed and demolished), Hangzhou, China

1991 Underground Entrances of Fengleqiao (completed and demolished), Hangzhou, China

1993 Lakeshore campus renovation planning (Scheme), China Academy of Art

1993 Building of Dept. of Sculpture (scheme), China Academy of Art

1994 Teaism garden of zen spirits (scheme), Hupao park, Hangzhou, China

1996-1999 Residence of a retired architecture professor (scheme)

1997 Self residence (interior), Hangzhou, China

1998 Chen Mo art studio (interior), Haining, China

1999-2000 Library of Wenzheng College, Suzhou University, Suzhou, China

1999-2000 Gallery with a View (completed and demolished), Shanghai, China

2000 Wall Gate (Instrument), Hangzhou, China

2001 Division (instrument), Hangzhou, China

2001-2005 Ningbo Contemporary Art Museum, Ningbo, China

2002-2003 Old town protection, antique architecture restoration and reconciliation Zone planning of Luqiao Town (scheme), Taizhou, China

2002-2004 Xiangshan Campus, China Academy of Art, Phase I, Hangzhou, China

2002-2007 Vertical Courtyard Apartments (Qianjiang Epoch), Hangzhou, China

2003-2004 慈城古建保护、古建修复及协调区（方案），中国宁波
2003-2005 东莞理工学院音乐与舞蹈系教学楼，中国东莞
2003-2006 瓷屋，中国金华
2003-2006 五散房，中国宁波
2003-2012 三合宅，中国南京
2004-2007 中国美术学院象山校区二期，中国杭州
2003-2008 宁波博物馆，中国宁波
2007-2009 中山路旧街区综合保护与更新，中国杭州
2008 华茂外国语学校华茂美术馆，中国宁波
2009 南宋御街博物馆，中国杭州
2009 和韵文化休闲中心（在建），中国昆明
2010 上海世博会宁波滕头馆，中国宁波
2010 金华城市文化中心（在建），中国金华
2010 宁海"十里红妆"传统嫁妆博物馆（在建），中国宁海
2010 船坞当代艺术博物馆（设计阶段），中国舟山渔港
2011 杭州佛学院图书馆（设计阶段），中国杭州
2011 瓦山：中国美术学院象山校区专家接待中心（在建），中国杭州

展览

1999 世界建筑师北京大会：中国青年建筑师实验建筑展，北京中国国际展览中心

2003-2004 Antique architecture protection, restoration and reconciliation zone, Planning of Cicheng Town (scheme), Ningbo, China
2003-2005 Teaching Building of Music and Dance Department, Dongguan Institute of Technology, Dongguan, China
2003-2006 Ceramic House, Jinhua, China
2003-2006 Five Scattered Houses, Ningbo, China
2003-2012 Sanhe House, Nanjing, China
2004-2007 Xiangshan Campus, China Academy of Art, Phase II, Hangzhou, China
2003-2008 Ningbo History Museum, Ningbo, China
2007-2009 Old Town Conservation and Intervention of Zhongshan Street, Hangzhou, China
2008 Huamao Art Museum, Huamao Foreign Language School, Ningbo, China
2009 Exhibition Hall of the Imperial Street of the Southern Song Dynasty, Hangzhou, China
2009 Heyun Culture and Leisure Center (under construction), Kunming, China
2010 Ningbo Tengtou Pavilion, Shanghai Expo, Ningbo, China
2010 City Cultural Center of Jinghua (under construction), Jinhua, China
2010 Ninghai "Shi Li Hong Zhuang" Traditional Dowry Musuem (under construction), Ninghai, China
2010 Contemporary Art Museum on the Dock (in design phase), Zhoushan, China
2011 Buddhist Institute Library of Hangzhou (in design phase), Hangzhou, China
2011 Tiles Hill – New Reception Center, Xiangshan Campus (under construction), Hangzhou, China

Exhibitions

1999 Chinese Young Architects Experimental Works, UIA Congress, Beijing, China

2000 杭州墙门：首届西湖国际雕塑邀请展
2001 杭州一分为二：第二届西湖国际雕塑邀请展
2001 变更通知：中国房子五人建造文献展
2001 土木：中国青年建筑展，德国柏林阿伊达斯美术馆
2002 都市营造：上海艺术双年展，上海美术馆
2003 第50届威尼斯双年展在广东美术馆、北京中央美院美术馆
2003 巴黎蓬皮杜艺术中心中国当代艺术展，巴黎
2004-2005 金华国际建筑艺术公园，中国国际小型公共建筑实践展，中国金华
2006 瓦园：威尼斯双年展第十届国际建筑展首届中国国家馆，威尼斯双年展
2006 "当代中国建筑展"荷兰鹿特丹国际建筑中心、荷兰建筑师协会
2007 中国建造：中国当代建筑展，纽约建筑中心
2007 香港国际建筑双城双年展，香港
2008 "梦想中国"当代中国建筑展：法国巴黎建筑与遗产城，夏悠宫
2008 活的中国园林：中国当代建筑与艺术展，德国德累斯顿博物馆
2009 "全球可持续建筑奖获奖建筑师作品展"，法国巴黎建筑与遗产城，夏悠宫
2009 "作为一种抵抗的建筑学"建筑个展，比利时布鲁塞尔艺术中心
2009 "M8急变中国／当代中国建筑展"，法兰克福德国建筑博物馆，柏林建筑中心
2010 "穹窿的塌朽"威尼斯双年展第十二届国际建筑展，威尼斯双年展

2000 Wall gate, 1st West Lake International Guest Exhibition of Sculptures, Hangzhou, China
2001 Division, 2nd West Lake International Guest Exhibition of Sculptures, Hangzhou, China
2001 Alteration Notice: Chinese House-construction Archive Exhibition of Five, Shanghai, China
2001 Tu Mu Young Architecture of China, AEDES Gallery, Berlin, Germany
2002 Shanghai Biennale, Shanghai Art Museum, Shanghai, China
2003 Synthi-Scapes: Chinese Pavilion of the 50th Venice Biennale, Guangdong Museum of Art, Guangzhou, and in Art Museum of Central Academy, Beijing, China
2003 Alors, La Chine, Centre Pompidou, Paris, France
2004-2005 Jinhau Architecture Park, Jinhua, China
2006 Tiled Garden – Chinese Pavilion of the 10th International Architecture Exhibition, Venice Biennale, Venice, Italy
2006 China Contemporary, Netherlands Architectural Institute (NAI), Rotterdam, The Netherlands
2007 Built in China – Architecture Exhibition, New York Architecture Center, USA
2007 Hong Kong Biennale of International Architecture, Hong Kong
2008 Dans la Ville Chinoise, Cité de l'Architecture et du Patrimoine, Palais de Chaillot, France
2008 Chinese Gardens for Living: From Illusion to Reality, Berg Palais, Dresden, Germany
2009 Exhibition of "Global Award for Sustainable Architecture 2007-8-9", Cité de l'Architecture et du Patrimoine, Palais de Chaillot, France
2009 "Architecture as Resistance" - solo exhibition, BOZAR Art Centre for Fine Arts, Brussels, Belgium
2009 "M8 in China, Deutsches Arkitectumuseum (DAM)", Frankfurt, Germany
2010 "Decay of a Dome", 12th International Architecture Exhibition, Venice Biennale, Venice, Italy

已发表中文文章及著作

文章

1987 "破碎背后的逻辑"南京工学院
王澍等人自办刊物
2001 "教育 / 简单"《时代建筑》
2003 "当'空间'开始出现"《建筑师》
2003 "走向虚构之城"《时代建筑》
2004 "同济记变"《时代建筑》
2004 "9＃咖啡室"《室内设计与装修》
2004 "小模型"《溢出的城市》
2005 "造房子的人"《图像建筑》
2005 "那一天"《时代建筑》
2005 "宁波美术馆设计手记"《建筑学报》
2005 "中国美术学院转塘校园设计"
《世界建筑》
2006 "'中国式住宅'的可能性"《时代建筑》
2006 "我们从中认出：宁波美术馆设计"
《时代建筑》
2006 "是建筑还是城市"《新设计》
2006 "当代大型城市建筑与地方性城市结构
的重建：宁波美术馆设计手记"《新设计》
2006 "中国美术学院转塘校园设计手记"
《新设计》
2006 "垂直院宅：杭州钱江时代"，
中国《世界建筑》
2007 "造园与造人"《建筑师》
2007 "观行之间"《建筑业导报》

Publications in Chinese

Articles

1987 "Logic behind the breaking", periodical edited by Wang Shu, etc., Nanjing Institute of Technology
2001 "Education/Simplicity", *Shi dai jian zhu = Time + Architecture*
2003 "when 'space' emerges", *Architects*
2003 "Heading to the Illusion City", *Shi dai jian zhu = Time + Architecture*
2004 "Record the Tongji Changes", *Shi dai jian zhu = Time + Architecture*
2004 "Cafe #9", *Interior Design Construction*
2004 "Small Models", *Overflew Cities*
2005 "The man who build", *Graphic Architecture*
2005 "One Day", *Shi dai jian zhu = Time + Architecture*
2005 "The Design of Ningbo Contemporary Art Museum", *Architecture Journal*
2005 "The Design of Xiangshan Campus, China Academy of Art", *World Architecture*
2006 "The Possibility of Chinese Style Dwelling Building", *Shi dai jian zhu = Time + Architecture*
2006 "We recognize: Notes on the Design of [the] Ningbo Museum of Art", *Shi dai jian zhu = Time + Architecture*
2006 "Building or City", *New Design*
2006 "The Reconstruction of Huge Urban Buildings and Local Urban Structure—Notes for the Creation of Ningbo Contemporary Art Museum", *New Design*
2006 "Notes for the Creation of Xiangshan Campus, China Academy of Art", *New Design*
2006 "Vertical Courtyard Apartments: Hangzhou Qianjiang Time, China", *World Architecture*
2007 "To Build a Garden and Educate a Person," *Architects*
2007 "Between Viewing and Doing," *Building Review*, Hong Kong

2008 "中国美术学院象山校园"《建筑学报》
2008 "中国美术学院象山校园二期工程"《时代建筑》
2008 "营造琐记"《建筑学报》
2008 "回想方塔园"《世界建筑导报》
2009 "自然形态的叙事与几何：宁波博物馆创作笔记"《时代建筑》
2009 "三合宅"《建筑技艺》
2010 "剖面的视野：宁波滕头案例馆"《时代建筑》
2010 "宁波五散房"《世界建筑导报》
2012 "循环建造的诗意：建造一个与自然相似的世界"《时代建筑》
2012 "我们需要一种重新进入自然的哲学"《世界建筑》

访谈

2006 "'中国式住宅'的可能性——王澍和他的研究生们的对话"《时代建筑》
2006 "'反学院'的建筑师：他的自称、他称和对话"《建筑师》
2006 "界于理论思辨和技术之间的营造"《建筑师》
2012 "问道：方振宁和王澍的对话"《艺术评论》
2012 "王澍访谈——恢复想象的中国建筑教育传统"《世界建筑》

著作

2002 "设计的开始"中国建筑工业出版社

2008 "Xiangshan Campus, China Academy of Art", *Architecture Journal*
2008 "Phase II of Xiangshan Campus, China Academy of Arts", *Shi dai jian zhu = Time + Architecture*
2008 "Construction Diaries", *Architecture Journal*
2008 "Recall Square Tower Garden", *World Architecture Review*
2009 "The narration and geometry of natural appearance: notes on the design of Ningbo Historical Museum", *Shi dai jian zhu = Time + Architecture*
2009 "Sanhe House", *Architecture Technique*
2010 "The field of vision on section", *Shi dai jian zhu =Time + Architecture*
2010 "Five Scattered Houses, Ningbo", *World Architecture Review*
2012 "Poetics of Construction with Recycled Materials: A World Resembling the Nature", *Shi dai jian zhu =Time + Architecture*
2012 "We are in Need of Reentering a Natural Philosophy", *World Architecture*

Interviews

2006 "Possibilities of 'Chinese-Style Housing': A Dialogue between WANG Shu and his Students", *Shi dai jian zhu =Time + Architecture*
2006 "The 'Anti-College' Architect", *Architects*
2006 "Constructions between the Theories and the Techniques", *Architects*
2012 "Questioning the way: A Dialogue between Fang Zhenning and Wang Shu", *Arts Criticism*
2012 "Interview with Wang Shu: Renew the Imaged Tradition of Chinese Architectural Education", *World Architecture*

Books

2002 "The Beginning of Design", China Architecture & Building Press

获奖情况

2004 首届中国建筑艺术奖

2005/2006 宁波五散房获瑞士 HOLCIM 基金会首届全球可持续建筑亚太区奖

2007 法国建筑师学会、法国建筑遗产城首届全球可持续建筑奖

2008 瑞士建筑奖提名

2008 杭州垂直院宅获德国法兰克福全球高层建筑奖提名

2010 作品"穹窿的塌朽"获第 12 届威尼斯双年展国际建筑展特别荣誉奖

2010 德国谢林建筑实践大奖

2011 法国建筑科学院金奖

2012 普利兹克建筑奖

Awards

2004 China's First Architecture Arts Award

2005/2006 Holcim Award for Sustainable Construction in the Asia Pacific for Five Scattered Houses in Ningbo

2007 First Global Award for Sustainable Architecture Cité de l'Architecture et du Patrimoine, France

2008 Nominated for the BSI Swiss Architecture Award

2008 Nominated for the German based International High Rise Award for the Vertical Courtyard Apartments, Hangzhou

2010 Special Mention for the Decay of a Dome exhibit 12th International Architecture Exhibit Venice, Italy

2010 Schelling Architecture Prize to Wang Shu and Lu Wenyu for significant designs, realized buildings or profound contributions to architectural history and theory

2011 Gold Medal from l'Academie d'Architecture of France

2012 Pritzker Architecture Prize

参考阅读

浙江美术学院国际画廊
《室内设计与装修》，1998（1）

穿越曲径迷园：王澍作品解读
《时代建筑》，2002（5）

文人建筑师的两副面孔
金秋野《建筑师》，2006，122

意义性的建筑解构：解读王澍的《那一天》及"中国美术学院象山新校园"
姜梅《新建筑》，2007（6）

内外山南：中国美术学院象山校园山南二期建筑访谈《时代建筑》，2008（3）

中国文人建筑传统现代复兴与发展之路上的王澍
赖德霖《建筑学报》，2012（5）

作为抵抗的建筑学——王澍和他的建筑
李翔宁《世界建筑》，2012（5）

读王澍和他的苏州文正学院图书馆
刘家琨《世界建筑》，2012（5）

形式书写与织体城市——作为方法和观念的象山校园 李凯生《世界建筑》，2012（5）

省略的世界——中国美术学院象山校园记述
葛明《世界建筑》，2012（5）

Reference Reading

"International Gallery of Zhejiang Academy of Fine Arts", *Interior Design Construction*, 1998(1)

"Wandering through a Meandering Garden: Dissecting the Works of Wang Shu", *Shi dai jian zhu = Time + Architecture*, 2002(5)

"Both Sides of literati Architect", by Jin Qiuye, *Architects*, 2006,122

"Meaningful Deconstruction to Architecture: Reading Wang Shu's Two Products–New Xiangshan Campus of the Chinese Academy of Art and One Day", by Jiang Mei, *New Architecture*, 2007(6)

"Interview about Phase II of Xiangshan Campus, China Academy of Arts", *Shi dai jian zhu = Time + Architecture*, 2008(3)

"Wang Shu: in the Context of the Revival and Development of Chinese Literati Architectural Tradition",by Lai Delin, *Architectural Journal*, 2012(5)

"Architecture as Resistance: Wang Shu and his Architecture", by Li Xiangning, *World Architecture*, 2012(5)

"About Wang Shu and the Library in Wenzheng College of Suzhou University", by Liu Jiakun, *World Architecture*, 2012(5)

"Calligraphy Form and Fabric City: Xiangshan Campus as Method and Concept",by Li Kaisheng, *World Architecture*, 2012(5)

"An Elliptical World: an Interpretation on Xiangshan Campus of Chinese Academy of Art", by Ge Ming, *World Architecture*, 2012(5)